To Mom for the inspiration, and for Stephen,
whose love and support never expire
—MC

This book is dedicated to all the new and existing
composters out there
—JC

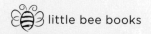 little bee books

New York, NY
Text copyright © 2022 by Melissa Coffey
Illustrations copyright © 2022 by Josh Cleland
All rights reserved, including the right of
reproduction in whole or in part in any form.
Manufactured in China   RRD 0522
First Edition   10 9 8 7 6 5 4 3 2 1
ISBN 978-1-4998-1254-1
Library of Congress Cataloging-in-Publication Data is available upon request.

littlebeebooks.com

For more information about special discounts on bulk purchases,
please contact Little Bee Books at sales@littlebeebooks.com.

FRIDGE-OPOLIS

Written by
Melissa Coffey

Illustrated by
Josh Cleland

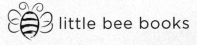 little bee books

Past Microwave Meadows and Dishwasher Downs
lay the infamous, polluted city of . . .

FRIDGE-OPOLIS

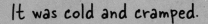

It was cold and cramped.

Dark and drippy.

Smelly as sauerkraut.

Mayor Mayonnaise knew he had a mess on his hands.

In the Midtown deli drawer,
the meats and cheeses were spoiling.

The hotdogs had all been forgotten.

Swiss cheese had turned moldy and bleu.

Italian salami caught whiffs of pastrami.

And the tofu was swimming in goo.

Downtown in the crispers, conditions were downright depressing.

Lettuce had long ago wilted.

Rhubarb was bitter and rude.

The overripe pineapple prickled.

Even broccoli was in a bad mood.

But more food kept moving into Fridge-opolis!

The condiments squeezed tighter in the Eastside high-rises.

Salsas had lost all their fire.

Dressings were cloudy with gloom.

Greek olives kept picking on pickles:

"Hey, dill juice—there ain't enough room!"

Even Uptown had turned into a jam-packed dump.

The butter was whipped in a frenzy.

Soy milk was curdling and sour.

Bagels and lox were ready to box.

Sushi felt mushy and dour.

All of Fridge-opolis had fallen into rancid ruin and disgusting decay.

And just when things couldn't get any worse . . .

Mayor Mayonnaise declared a state of emergency.

He sent an envoy to Undersink Labs.

Only one thing could save them . . .

Doctor Baking Soda heeded
the Mayor's desperate call.

"We'll scrub every germ from your city,

hose down every
borough and street.

After that fateful day, Fridge-opolis citizens rejoiced.
The Mayor and Doctor Baking Soda were heroes!

Doctor Baking Soda was given the key to the city.
Order was restored.

The jello had plenty of jiggle.

The orange juice never felt squeezed.

The bacon and eggs didn't scramble.

The peas in their pods were quite pleased.

The turkey was not acting jerky.

The ketchup and mustard made peace.

The apples quit being so crabby.

All the stinkiness finally did cease.

Now, past Recycling Ridge and Compost Town,
lies the famous, sparkling city of Fridge-opolis . . .

where everyone's happy as pie.

FOOD FOR THOUGHT

Do you have a Fridge-opolis in your kitchen? This story is a playful way of introducing the concept of food waste, but it is indeed quite a serious problem. In the United States, we waste up to a whopping 40 percent of all food each year! From lunch box leftovers to brown bananas, the little bites we don't eat add up. Every year, each of us throws out about 290 pounds of food (the size of a baby elephant). Yet, up to fifty million Americans—many of them kids—still go hungry.

When food gets dumped, it wastes the resources needed to make it: water, land, energy, and a lot of money ($218 billion). It also generates harmful greenhouse gases, which cause climate change. Food waste is a really huge problem, but one that everyone can help solve! By changing our habits and doing small things every day, we can make a big difference. If all of us do our part, we can reach our national goal of cutting food waste and loss in half by 2030.

Doctor Baking Soda and Mayor Mayonnaise Food Hero Tips

Make a shopping list
Help plan meals
Buy funny-looking (ugly) produce
Store food properly
Embrace your freezer
Understand expiration dates
Give leftovers some love
Take right-size portions
Share or donate

WHAT A WASTE

Never eaten:

45% of fruits/veggies
35% of fish/seafood
30% of grains
20% of meat
20% of dairy

Loaded Landfills

Food is the number-one thing dumped in America's landfills. When it rots, it pollutes our climate with a greenhouse gas called methane, which is eighty-six times more powerful than carbon dioxide at warming the Earth.

Compost Town

Composting is nature's way of recycling. Do a "heap" of good and make your own bin of kitchen scraps, dirt, and water. Brown + green = magic fertilizer! Got worms? You can make a worm bin (vermicompost) indoors or out with just a small bin, shredded paper, and some wiggly friends. (Worm poop is great for the soil!)

Recycling Ridge Rocks

Think twice before tossing and look for the symbol! One aluminum can will sit in a landfill anywhere between eighty to five hundred years. Rinse and recycle glass, plastic, and cardboard to help our planet go green.

A soda can can save enough energy to watch tv for three hours.

A glass bottle can power a light bulb for four hours.

Select sources:
www.savethefood.com
www.foodprint.org/issues/the-problem-of-food-waste
www.feedingamerica.org
www.usda.gov/foodlossandwaste/why
www.refed.com
www.cec.org/flwy
Gifford, Clive. *Guardians of the Planet: How to be an Eco-Hero.* B.E.S. Publishing, 2020.